# LOW ENERGY TRANSMUTATION OF NUCLEAR WASTE

Healing the Earth for Future Generations

# EDWARD ESKO

IMI Press
LENOX, MA

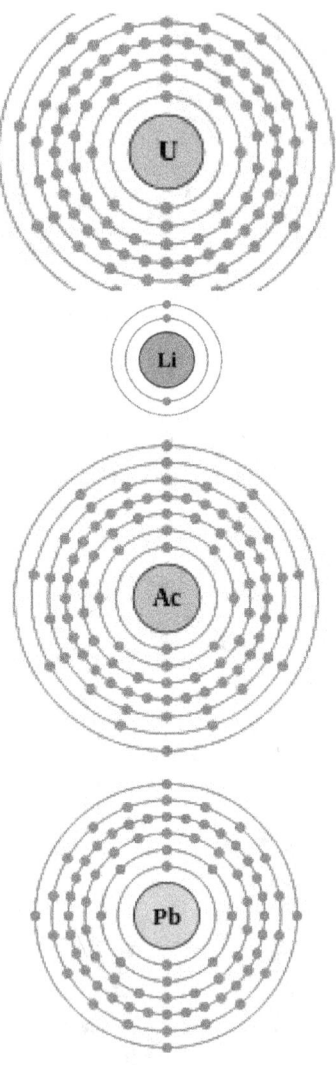

The low energy fission of $^{235}$U into $^{228}$Ac could shorten the half life of uranium-235 from 700 million years to 1.9 years

Separate ripples amplify and merge into larger ripples. In low energy transmutation nuclei may undergo a similar transformation

The giant casks of spent nuclear fuel loom over the future like a toxic Stonehenge. They bear silent witness to our inability to solve the problem of nuclear waste. They are the sentinels of our ignorance.

# LOW ENERGY TRANSMUTATION OF NUCLEAR WASTE

## CONTENTS

Introduction   7
Low Energy Transmutation of Nuclear Waste   13

    Uranium-235   14
    Plutonium-239   17
    Radium-226   18
    Cesium-137   19
    Iodine-129   22
    Technitium-99   23
    Guidelines for Methodology   24
    Conclusion   29

Interview with Edward Esko   35

Low Energy Transmutation of Nuclear Waste
Reprinted from *Infinite Energy* magazine
Diagrams by Steve Hansen

Copyright © 2012/2019 by Edward Esko
ISBN-13: 978-1729588734
ISBN-10: 1729588735

Published by IMI Press
A division of the International Macrobiotic Institute (IMI)
P.O. Box 2051, Lenox, MA 01240
(413) 446-2620
QuantumResearchInstitute.com
edwardesko@gmail.com

Much progress has taken place in establishing firmly the occurrence of different types of transmutation reactions in a wide variety of configurations.

Mahadeva Srinivasan<br>Bhabha Atomic Research Center (Retired)

# INTRODUCTION

> The problem is how to keep radioactive waste in storage until it decays after hundreds of thousands of years. The geologic deposit must be absolutely reliable as the quantities of poison are tremendous. It is very difficult to satisfy these requirements for the simple reason that we have had no practical experience with such a long-term project.
>
> Hannes Alfven, Nobel laureate in physics

Once each season for several years the cool fusion research team would meet at the Moore Mill building in Bellows Falls, Vermont, to conduct tabletop carbon-arc experiments. Alex Jack and I would drive from Western Massachusetts and meet up with Woody Johnson. The drive took us east on the Mass Pike and then north up route 91 past Brattleboro to the Rockingham exit. Bellows Falls is on the Connecticut River. Thirty miles downriver, in the town of Vernon, is the Vermont Yankee nuclear power plant.

Vermont Yankee epitomizes the chaotic state of affairs within the nuclear industry. The amount of radioactive waste stored at Vermont Yankee is substantial. It exceeds that of all four damaged reactors at Fukushima in Japan. The used fuel rods at Vermont Yankee are about one million times more radioactive than they were before being used in the reactor.

The rods are hot enough to catch fire if they are not stored under water. Five hundred tons of spent fuel is now being stored in pools of water seven stories above ground.

Following decades of conflict with local residents and with the State of Vermont, Vermont Yankee closed in 2014. The Yankee plant went on line in 1972 and is the same General Electric boiling water system as the failed nuclear reactors at Fukushima. Following the disaster at Fukushima, Entergy, the owner of Vermont Yankee, was being pressured to make expensive modifications to improve safety.

With the closing of the plant, Vermont Yankee stopped producing radioactive waste. However, still unresolved is what will happen with the tons of highly radioactive spent fuel rods stored at the site. Local residents are justifiably concerned. At the time of the shutdown, one resident, the director of a local citizen advocacy group, stated, "There are at least 530 tons of high-level (radioactive) waste, which we've said still needs to come out of that spent fuel pool. This isn't over. The struggle is now about cleanup."

Another resident stated, "The banks of the river, within a 500-year flood plain, is not the best place to store high-level radioactive waste for even a short period of time. Unfortunately, there is no long-term storage in the U.S., and we're probably stuck with it there, just like at Yankee Rowe." Yankee Rowe is the former nuclear plant in nearby Rowe, Massachusetts that was shut down in 1992. High-level radioactive waste from Yankee Rowe is now in temporary storage in 16 dry casks at the site. The giant casks of spent nuclear fuel loom over the future like a modern, highly toxic Stonehenge. They bear silent witness to our inability, at least for now, to solve the problem of nuclear waste. They are the sentinels of our ignorance. When the shutdown was announced, a third resident expressed concern about the safety of Vermont Yankee's phase-out: "Now until the fall of 2014 will be the most dangerous year of their operation of the plant, because the plant will be older than ever, parts will be more brittle than ever, they will be more reluctant than ever to repair and replace parts … and workers who have already been let go are going to be leaving in much larger numbers. To expect that Entergy is going to be taking every single precaution to keep the plant as safe as it possibly can be is unfortunately unrealistic."

Prior to the shutdown, a member of the citizen advocacy group, the New England Coalition, explained the situation at Vermont Yankee quite succinctly:

"One fundamental purpose of our advocacy has always been to try to protect the public and the environment from nuclear waste—waste in the fuel, in the reactor, in the pool, out-in-the-yard, soon to be released in the next reactor or fuel handling accident, and out on the wind. Soon, Entergy Vermont Yankee, a nuclear waste pile that generated electricity will stop generating electricity—and it will either be mothballed or promptly torn apart, but it will be, absent electricity generation, just a nuclear waste pile … from which the public and the environment need to be protected."

On a planetary scale, there are more than 430 locations around the world where nuclear waste continues to accumulate. Most is stored at individual reactor sites. Nuclear reactors on planet earth create about 10,000 metric tons of spent nuclear fuel each year. Thus, the problem gets worse with each passing day. The shutdown and decommission of nuclear power plants solves only the problem of new nuclear waste. It does nothing to solve the problem of already existing waste. Moreover, the process is expensive (between $300 million to $5.6 billion per unit), time-consuming, and hazardous to workers and the natural environment. It opens a window for disaster caused by human error, accident, or sabotage. In the U.S. there are 13 reactors that have shut down and are in the process of decommission. None have fully completed the process that can last as long as 100 years.

The timeframes when dealing with nuclear waste are enormous; they range from 10,000 years to millions of years. Storage of nuclear waste, whether "temporary" or "permanent," does not solve the problem. It simply passes it on to future generations. The very existence of nuclear waste is itself the problem. For the sake of future generations, we need to seriously investigate promising ideas not just for storing nuclear waste, but also for actually *getting rid* of it.

Toward that end, Quantum Rabbit LLC has been conducting tabletop research on the low energy transmutation. Transmutation is defined as transforming one atom into another by changing its nuclear structure.

Low energy transmutation attempts to achieve this with simple tabletop equipment, using electric power from car batteries, solar panels, a HUBERT® portable generator, and the wall socket, with relatively low temperatures and low pressures created in simple glass vacuum tubes. This is in contrast to the high-energy accelerator transmutation of waste (ATW) being studied around the world as a possible solution to the radioactive waste problem.

In theory, the process of subtracting lithium, boron, carbon, or another light element from lead and other super-heavy elements could be used to condense the decay cycle of radioactive elements such as uranium-235 and plutonium-239 from thousands or millions of years to a few years at most.

On a parallel track, low energy fusion could also be deployed to instantly convert radioactive fission products like iodine-129, technetium-99, and cesium-137 into useful elements like barium, palladium, and neodymium. Clearly, much work needs to be done. We plan to continue experiments at our small labs in New England, but with limited time and resources, our work remains preliminary, like that of Steve Jobs and Steve Wozniak in their garage in Silicon Valley or the Wright Brothers in their bicycle shop in Ohio. Our hope is to eventually partner with established labs or universities to take our preliminary work to the next level.

Nuclear energy was unleashed seventy years ago with the Manhattan Project. In today's dollars, the creation of nuclear power cost about $26-billion. With research and development in 30 locations, the project employed 130,000 people. It may require a similar effort today, with researchers around the globe committed to the same goal, to put the nuclear genie back into the bottle. We owe it to future generations to act now to solve this seemingly intractable problem that threatens humanity and planet earth.

Edward Esko
Pittsfield and Lenox, Massachusetts

# LOW ENERGY TRANSMUTATION OF NUCLEAR WASTE

Quantum Rabbit (QR) research on the low energy fusion and fission (low energy transmutation; also known as low energy nuclear reactions, or LENR) of various elements indicates possible pathways for applying that process to reducing nuclear materials. In New Energy Foundation (NEF)-funded tests conducted at the Quantum Rabbit labs in Nashua, NH and Owls Head, ME, QR researchers initiated a possible low energy fission reaction in which lead-204 fissioned into lithium and gold ($^{204}Pb \rightarrow {}^7Li + {}^{197}Au$). [1] This reaction may have been triggered by a low energy fusion reaction in which lithium fused with sulfur to form potassium ($^7Li + {}^{32}S \rightarrow {}^{39}K$).

These results confirm earlier findings showing apparent low energy fusion and fission reactions. [2] Moreover, subsequent research with boron indicates apparent low energy reactions in which boron fuses with oxygen to form aluminum and with sulfur to form scandium. [3]

At the same time, the QR group has achieved what appear to be low energy transmutations of carbon using carbon-arc under vacuum and in open air. [4] The research group at QR believes these processes can be adapted to accelerate the natural decay cycle of uranium-235, plutonium-239, radium-226, and the fission products cesium-137, iodine-129, and technetium-99, with the long-term potential of reducing the threat posed by radioactive isotopes to human health and the environment.

**Uranium-235**

Radioactive uranium is the primary constituent of spent nuclear fuel. The half-life of uranium-235 is more than 700 million years. The first step in this process, the alpha decay of uranium-235 into thorium-231 consumes the bulk of this enormous span. The half-lives of the isotopes that follow thorium-231 total approximately 33,000 years with the stable isotope lead-207 as the conclusion of the process.

The QR research indicates it may be possible to intervene in the decay cycle of uranium in order to reduce the amount of time needed to achieve its transmutation into lead. The most obvious window for intervention is at the beginning of the cycle, by inducing uranium-235 to fission into one of the lighter isotopes in the radioactive decay chain.

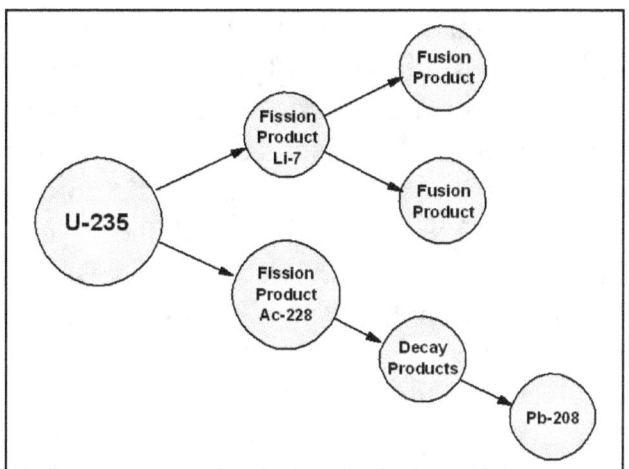

Fig 1. Low energy transmutation of uranium-235

We propose using lithium, the catalyst element in the studies cited above, as the catalyst for the following low energy fission reaction:

$^{235}U \rightarrow \,^{7}Li + \,^{228}Ac$
Uranium-235 → lithium-7 + actinium-228

According to this hypothesis, the low energy fusion of lithium with sulfur, resulting in potassium, triggers the low energy fission of uranium into lithium and actinium. The low energy fusion reaction can be written as follows:

$^{7}Li + \,^{32}S \rightarrow \,^{39}K$
Lithium-7 + sulfur-32 → potassium-39

These reactions are summarized in Fig. 1. If achieved, they set in motion the natural decay cycle beginning with actinium-228 and ending with lead-208 shown in Fig. 2. Note that the low energy transmutation that causes the uranium-235 to fission into actinium-228 results in U-235 being cycled downstream into the natural decay chain of thorium-232. [5] If actinium-228 is produced as predicted, and the natural decay-cycle indicated in Fig. 2 set in motion, the half-life of uranium-235 is compressed from over 700-million years to slightly over 1.9 years. The process is summarized in the formula:

$$^{235}U \rightarrow {}^{7}Li + {}^{228}Ac \rightarrow {}^{208}Pb$$

Uranium-235 → lithium-7 + actinium-228 (thorium-232 decay cycle) → lead-208

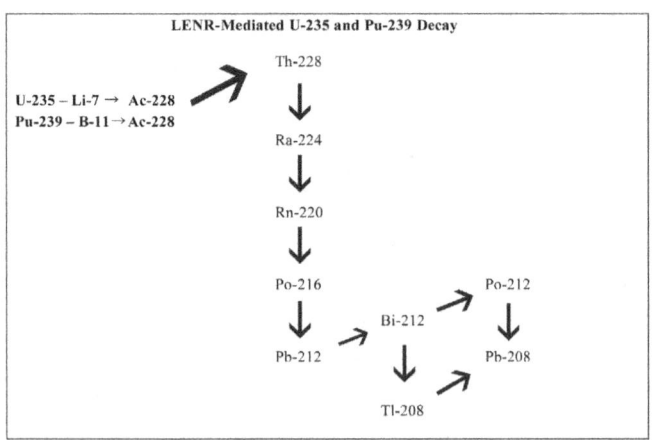

Fig. 2. Accelerated decay series: U-235 and Pu-239. Downward arrows represent alpha decay; upward arrows beta decay

**Plutonium-239**

There is a significant quantity of deadly plutonium-239 in spent nuclear fuel. Plutonium-239 has a half-life of 24,000 years. The QR research group achieved promising results with the low energy fusion of boron. These tests are outlined in the book, *Cool Fusion*. A series of experiments for the possible reduction of plutonium-239 similar to the QR boron experiments can be designed using boron as the catalyst element. The low energy fission reaction we propose testing is as follows:

$^{239}Pu \rightarrow {}^{11}B + {}^{228}Ac$
Plutonium-239 → boron-11 + actinium-228

This low energy fission reaction is theoretically triggered by several low energy fusion reactions: [6]

$^{11}B + {}^{16}O \rightarrow {}^{27}Al$
Boron-11 + oxygen-16 → aluminum-27

$^{11}B + {}^{34}S \rightarrow {}^{45}Sc$
Boron-11 + sulfur-34 → scandium-45

Once again, if low energy transmutation is successful in producing actinium-228, like U-235 in the formula described above, Pu-239 will be cycled downstream into the thorium-232 decay chain with the end product being the stable isotope lead-208 (Fig. 2). This process can be summarized as follows:

$^{239}$Pu → $^{11}$B + $^{228}$Ac → $^{208}$Pb
Plutonium-239 → boron-11 + actinium-228 (thorium-232 decay cycle) → lead-208

### Radium-226

Contamination by radium-226 continues to be a problem at U.S. military installations and other sites around the world. Radium-226 is part of the U-238 decay chain with a half-life of 1,600 years. With low energy transmutation, it may be possible to compress this time frame considerably by achieving the low energy fission of Ra-226.

QR research on carbon-arc may offer a method for achieving this possibility. Numerous low energy transmutations have been reported, both in open air and under vacuum. [7] These low energy fusion reactions could possibly be used to prompt the low energy fission of radium-226, compressing the half-life of radium and accelerating the natural decay cycle from more than 1,600 years to approximately 22 years. (Fig. 3.)

The low energy fission reaction we propose testing is as follows:

$^{226}$Ra → $^{12}$C + $^{214}$Pb
Radium-226 → carbon-12 + lead-214

This low energy fission reaction could possibly be triggered by low energy fusion reactions such as those between carbon and oxygen noted in QR carbon-arc research:

$^{12}C + {}^{12}C \rightarrow {}^{24}Mg$
Carbon-12 + carbon-12 → magnesium-24

$^{12}C + {}^{16}O \rightarrow {}^{28}Si$
Carbon-12 + oxygen-16 → silicon-28

$^{12}C + 2({}^{16}O) \rightarrow {}^{44}Ti$
Carbon-12 + 2(oxgyen-16) → titanium-44

$^{12}C + {}^{32}S \rightarrow {}^{44}Ti$
Carbon-12 + sulfur-32 → titanium-44

$2({}^{12}C + {}^{16}O) \rightarrow {}^{56}Fe\ ^{(+2\ protons)}$
2(Carbon-12 + oxygen-16) → iron-56 + two protons

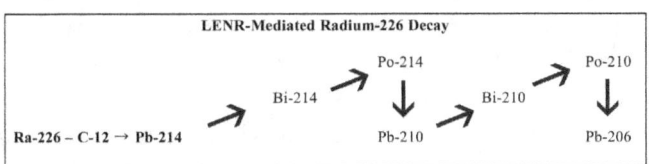

Fig. 3. Accelerated decay series: Ra-226. Downward arrows represent alpha decay; upward arrows beta decay

## Cesium-137

Cesium-137, a product of nuclear fission is a major radionuclide in spent nuclear fuel. It has a half-life of 30 years and decays by emitting a beta particle. Its decay product, barium-137m (the "m" is for metastable) stabilizes by emitting an energetic gamma ray with a half-life of approximately 2.6 minutes. It is this decay product that qualifies cesium-137 as a radiation hazard.

The environmental dangers posed by cesium-137 were highlighted by the crisis at Fukushima Daichi reactor in Japan. Writing in the Proceedings of the National Academy of Sciences, [8] an international team of scientists described the threat posed by cesium-137:

> The largest concern on the cesium-137 ($^{137}$Cs) deposition and its soil contamination due to the emission from the Fukushima Daiichi Nuclear Power Plant (NPP) showed up after a massive quake on March 11, 2011. Cesium-137 ($^{137}$Cs) with a half-life of 30.1 y causes the largest concerns because of its deleterious effect on agriculture and stock farming, and, thus, human life for decades. Removal of $^{137}$Cs contaminated soils or land use limitations in areas where removal is not possible is, therefore, an urgent issue.

Contamination by cesium-137 was a major problem following the Chernobyl disaster. As John Emsley states: [9]

> Uranium fuel rods in nuclear power stations produce cesium-137. The half-life of cesium-137 is 30 years, which means that it takes over 200 years to reduce it to 1% of its former level. For this reason, an accident at a nuclear power plant can contaminate the environment around for generations, which is why the Chernobyl accident in the Ukraine in 1986 was such an environmental disaster. It released a large amount of radioactive cesium-137 which drifted all over Western Europe, affecting sheep farms as far west as Scotland, Ireland, and Wales, over 1500 miles from the accident.

There it was washed to earth by heavy rain and taken up by the roots of plants, thus becoming part of the vegetation that sheep ate.

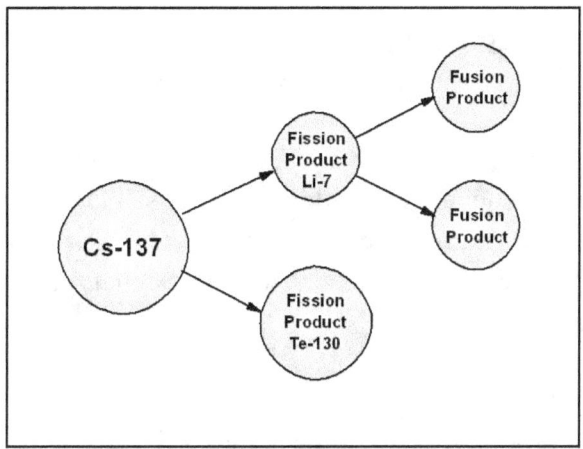

Fig. 4. Low energy transmutation of cesium-137

Using low energy transmutation, it may be possible to convert cesium-137 to tellurium-130, a stable non-radioactive isotope, thus redirecting and compressing the cesium-137 decay cycle (Fig. 4). The low energy fission formula is as follows:

$^{137}Cs \rightarrow {}^{7}Li + {}^{130}Te$
Cesium-137 → lithium-7 + tellurium-130

In theory, the low energy fission reaction would be triggered by the low energy fusion of lithium and sulfur:

$^{7}Li + {}^{32}S \rightarrow {}^{39}K$
Lithium-7 + sulfur-32 → potassium-39

In a separate experiment, cesium-137 may also transmute into neodymium-148 through a low energy fusion reaction:

$^{137}Cs + {}^{11}B \rightarrow {}^{148}Nd$
Cesium-137 + boron-11 → neodymium-148

If the fusion reaction can be proven and scaled to production levels, it would then be possible to convert dangerous radioactive waste into a valuable rare earth metal widely utilized today in the magnets in hybrid vehicles.

### Iodine-129

Iodine-129 is a long-lived isotope of iodine created primarily from the fission of uranium and plutonium in nuclear reactors. It decays with a half-life of 15.7 million years.

Significant amounts of iodine-129 were released into the atmosphere following nuclear weapons tests in the 1950s and 1960s. Iodine-129 is long-lived and mobile in the environment and is thus of special importance in disposal and management of spent nuclear fuel.

It may be possible to compress the natural decay cycle of this radioisotope through the process of low energy fission. The induced fission reaction is as follows:

$^{129}I \rightarrow {}^{7}Li + {}^{122}Sn$
Iodine-129 → lithium-7 + tin-122

Once again, according to theory, low energy fusion of lithium and sulfur would serve as the catalyst for the reaction:

$^7Li + {}^{32}S \rightarrow {}^{39}K$
Lithium-7 + sulfur-32 → potassium-39

During the experiment, iodine-129 may also transmute into barium-146 through a simultaneous fission reaction:

$^{129}I + {}^7Li \rightarrow {}^{136}Ba$
Iodine-129 + lithium-7 → barium-136

### Technetium-99

Technetium-99 is radioisotope of technetium that decays with a half-life of 211,000 years to stable ruthenium-99. It is the most significant long-lived fission product of uranium-235. Its high fission yield, relatively long half-life, and mobility in the environment make technetium-99 one of the more problematic components of nuclear waste.

There have been releases into the environment from atmospheric nuclear tests, nuclear reactors, and in the late 1990s from the Sellafield plant, which released nearly 1,000 kg into the Irish Sea. It may be possible to accelerate the half–life of Tc-99 by inducing the following low energy fission reaction:

$^{99}Tc \rightarrow {}^7Li + {}^{92}Zr$
Technetium-99 → lithium-7 + zirconium-92

Once again, in theory, the reaction would be triggered by the low energy fusion of lithium and sulfur:

$^{7}Li + {}^{32}S \rightarrow {}^{39}K$
Lithium-7 + sulfur-32 → potassium-39

During the experiment, Tc-99 may also transmute into Pd-106 through the following fusion reaction:

$^{99}Tc + {}^{7}Li \rightarrow {}^{106}Pd$
Technetium-99 + lithium-7 → palladium-106

**Guidelines for Methodology**

The experiments on low energy transmutation cited in *Cool Fusion* and *Corking the Nuclear Genie* can serve as a starting point for designing experiments to test the nuclear reduction hypothesis presented in this paper. [10] Vacuum tubes similar to those used in the QR low energy transmutation tests and shown in Fig. 5 can be considered for the nuclear reduction tests. Because silver is a strong conductor of electricity and a neutron absorber, we propose using it as the anode and cathode material, with other test materials adjusted for each experiment as indicated below (Fig. 6).

Moreover, silver may react independently with lithium to form tin ($^{109}Ag + {}^{7}Li \rightarrow {}^{116}Sn$). This reaction was noted in a previous QR test. [11]

Keep in mind that these suggestions are guidelines only, based on previous low energy fusion and fission experiments.

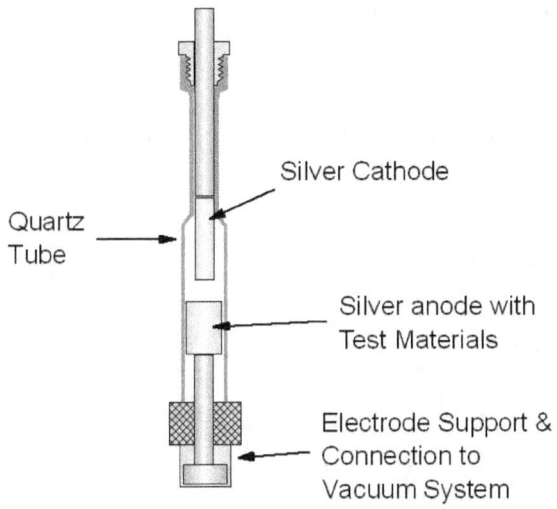

Fig. 5. Tube and electrode configuration

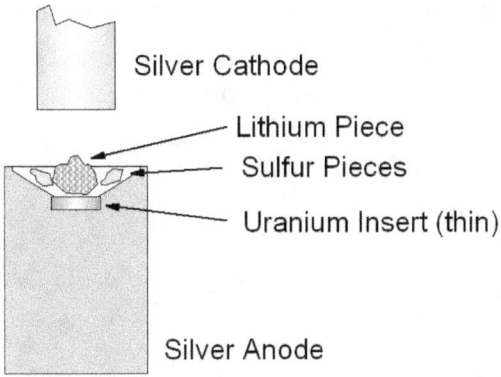

Fig. 6. Electrodes and test material suggested for the U-235 → Li-7 + Ac-228 experiment

***Uraninum-235:***

$^{235}U \rightarrow {}^7Li + {}^{228}Ac \rightarrow {}^{208}Pb$

Electrodes made of Ag
Test Materials:

1. Uranium insert (thin wafer or foil) in anode
2. Lithium test material
3. Sulfur test material
4. Pure neon/oxygen backfill

***Plutonium-239:***

$^{239}Pu \rightarrow {}^{11}B + {}^{228}Ac \rightarrow {}^{208}Pb$

Electrodes made of Ag
Test Materials:

1. Plutonium insert (thin wafer or foil) in anode
2. Boron test material
3. Sulfur test material (optional)
4. Pure neon/oxygen backfill

***Radium-226:***

$^{226}Ra \rightarrow {}^{12}C \rightarrow {}^{214}Pb \rightarrow {}^{206}Pb$

Electrodes made of Ag
Test Materials:

1. Radium insert (thin wafer or foil) in anode
2. Carbon (graphite) test material
3. Sulfur test material
4. Pure nitrogen/oxygen backfill*

*Note: Adding nitrogen allows the process to take advantage of potential carbon-nitrogen reactions such as those noted in QR research. [12]

***Cesium-137:***

$^{137}Cs \rightarrow {}^{7}Li + {}^{130}Te$

Electrodes made of Ag
Test Materials:

1. Cesium insert (thin wafer or foil) in anode
2. Lithium test material
3. Sulfur test material
4. Pure neon/oxygen backfill

$^{137}Cs + {}^{11}B \rightarrow {}^{148}Nd$

Electrodes made of Ag
Test Materials:

1. Cesium insert (thin wafer or foil) in anode
2. Boron test material
3. Sulfur test material
4. Pure neon/oxygen backfill

### *Iodine-129:*

$$^{129}I \to {}^7Li + {}^{122}Sn$$
$$^{129}I + {}^7Li \to {}^{136}Ba$$

Electrodes made of Ag
Test Materials:

1. Iodine inserted in or on anode
2. Lithium test material
3. Sulfur test material
4. Pure neon/oxygen backfill

### *Technetium-99:*

$$^{99}Tc \to {}^7Li + {}^{92}Zr$$
$$^{99}Tc + {}^7Li \to {}^{106}Pd$$

Electrodes made of Ag
Test Materials:

1. Technetium insert (thin wafer or foil) in anode
2. Lithium test material
3. Sulfur test material
4. Pure neon/oxygen backfill

### *Procedure for the Above Experiments:*

1. Insert is placed on or into the anode.
2. Measured quantity of test materials are placed in anode recess.

3. Glass/quartz tube is placed over the anode assembly.
4. Cathode is inserted into the tube and secured at the desired separation from the anode.
5. Fill with neon (or nitrogen for Ra-226) to 2 torr.
6. Strike plasma using direct current (DC).
7. Admit oxygen fill to 6 torr. Continue until reaction noticeably slows or tube is in danger of breaking (approximately 10-20 minutes.)
8. Disconnect power and allow sample to cool.

**Conclusion**

As of this writing, the problem of nuclear waste disposal remains unsolved. In an op-ed published in the *Santa Monica Daily Press,* [13] Dr. Jeffrey Patterson, former head of Physicians for Social Responsibility (PSR) stated:

> 2011 was a scary year for nuclear reactor sites. The summer floods threatened to encroach on reactors in Nebraska and Iowa, an earthquake and a hurricane happened in quick succession to rattle and flood the East Coast, and the continuing events of the Fukushima-Daichi reactor accident provided harrowing examples of the threats posed to spent fuel at reactor sites.
>
> The fate of spent fuel there kept the world on edge for days. It's worth noting that the amount of fuel in vulnerable storage pools in Japan was far less than what is crowded into pools at many U.S. reactors. As we all learned, a loss of coolant could produce a fuel melt and large radiation release.

It wasn't supposed to be this way. Used reactor fuel was to be permanently stored in deep underground repositories, away from floods and other natural hazards. But the solution to the nation's nuclear waste problem has been elusive for decades. Meanwhile, 65,000 metric tons of spent reactor fuel is still looking for a home.

The Blue Ribbon Commission on America's Nuclear Future proposes transferring spent nuclear fuel, now scattered at 70 locations around the U.S., to temporary storage areas, pending selection of more permanent deep geologic repositories. This proposal is not without controversy. As Dr. Patterson states:

> Moving spent fuel around the country is not a risk worth taking. Rather than addressing the problem, an "interim" facility would only relocate it. So what is the best option? Hardened on-site storage of spent fuel. It's safe, cost-effective—and readily available. PSR and over 170 public interest organizations from all 50 states are calling for adoption of this approach.
>
> Storing reactor fuel at reactor sites in hardened buildings that can resist severe attacks, such as a direct hit by high-powered explosives or a large aircraft, as is done in Germany, offers the safest and most sensible option until a permanent repository can be found.

These proposals offer opportunities for research on low energy transmutation. Research laboratories could be set up at future on-site hardened facilities or even now at current waste storage sites, as well as at future interim facilities.

These laboratories can begin first-round investigation of low energy transmutation. If successful, scale-up can proceed to levels required to reduce the on-site, regional, and global inventory of nuclear waste. Moreover, low energy transmutation may offer an efficient low-cost alternative to accelerator transmutation of waste (ATW). In 1999, the U.S. Department of Energy's (DOE) Office of Civilian Radioactive Waste Management submitted a report to Congress entitled "A Roadmap for Developing Accelerator Transmutation of Waste (ATW) Technology." Sekazi K. Mtingwa of MIT describes this approach as follows: [14]

> Transmutation means the transformation of one atom into another by changing its nuclear structure. In the present context this means bombarding a highly radioactive atom with neutrons, preferably fast neutrons, from either a fast nuclear reactor or spallation neutrons created by bombarding protons from a high-energy accelerator on a suitable target.

The Oak Ridge National Laboratory (ORNL) is currently investigating methods for accelerator transmutation (ATW) of nuclear wastes. An article in the ORNL *Review* states: [15]

> Conceived by scientists at Los Alamos National Laboratory, ATW uses a linear accelerator system to produce neutrons for transmutation of excess weapons plutonium and other radioactive DOE wastes, such as technetium-99 and iodine-129.

> Ultimately, the potential of partitioning and transmutation to waste management is this: If a radioactive waste stream no longer exists, then it poses no radiological hazard. More than anything else, this simple fact has spurred the recent resurgence of interest in partitioning-transmutation technology.

Meanwhile, in the Eurozone, the European nuclear establishment is pressing ahead with a $1.2 billion R&D project to look into high-energy neutron-induced transmutation.

The first stage of the project, the setup of a demonstration system known as "Guinevere" that combines a particle accelerator and a nuclear reactor, took place in January 2012 at the Belgian Nuclear Research Center at Mol. A larger version of the reactor system, known as Myrrha (Multipurpose Hybrid Research Reactor for High-tech Applications), is scheduled to become operational in 2023. A press release from the World Nuclear Association explains the thinking behind the project: [16]

> Myrrha will be able to produce radioisotopes and doped silicon, but its research functions would be particularly well suited to investigating transmutation. This is when certain radioactive isotopes with long half-lives are made to "catch" a neutron and thereby change into a different isotope that will decay more quickly to a stable form with no radioactivity. If achievable on an industrial scale, transmutation could greatly simplify the permanent geologic disposal of radioactive waste.

The Quantum Rabbit group estimates that research on low energy transmutation could begin at a fraction of the estimated $1.2 billion startup cost of the Myrrha project. (QR estimates $1.2 million for feasibility study and $12 million to develop a prototype system, amounts that are respectively 0.1% and 1% the cost of Myrrha.) Rather than a highly centralized billion-dollar processing system, low energy transmutation technology could be distributed to nuclear power stations around the globe at an affordable cost.

The task of nuclear remediation would become the responsibility of the individual power station and thus remain local instead of becoming highly centralized. Also, the amount of power needed to conduct low energy transmutation would be miniscule compared to the power required to operate a particle accelerator and nuclear reactor. At the very least, research on low energy transmutation should proceed on a parallel track to the high-energy neutron-induced transmutation projects currently underway or under consideration in order to determine which approach yields the most promising results.

[1] Esko, Edward, "Anomalous Metals Part II," *Infinite Energy*, No. 103, 2012.
[2] Esko, Edward and Jack, Alex, *Cool Fusion*, second edition, Amber Waves, Becket, Mass., USA, pp. 56-113, 132-151, 2012.
[3] Esko, Edward, "In Search of the Platinum Group Metals Part II," *Infinite Energy*, No. 104, 2012.
[4] Esko, Edward and Jack, Alex, *Cool Fusion*, second edition, Amber Waves, Becket, Mass., USA, pp. 56-60, 88-97, 2012.

[5] Argonne National Laboratory, "Human Health Fact Sheet," Fig. N.3 Natural Decay Series: Thorium-232, 2005.

[6] Esko, Edward, "In Search of the Platinum Group Metals Part II," *Infinite Energy*, no. 104, 2012.

[7] Esko, Edward and Jack, Alex, *Cool Fusion*, second edition, Amber Waves, Becket, Mass., USA, pp. 56-60, 88-97, 2012.

[8] Yasunari, Teppei J., Stohl, Andreas, Hayano, Ryugo S., Burkhart, John F., Eckhardt, Sabine, Yasunari, Tetsuzo, "Cesium-137 deposition and contamination of Japanese soils due to the Fukushima nuclear accident," Proceedings of the National Academy of Sciences, November, 14, 2011.

[9] Emsley, John, *Nature's Building Blocks: An A-Z Guide to the Elements*, first edition, Oxford University Press, Oxford, England, pp. 82, 2001.

[10] Refer to Occupational Safety and Health Administration (OSHA) guidelines for the handling of hazardous materials prior to initiating these experiments.

[11] Esko, Edward and Jack, Alex, *Cool Fusion*, second edition, Amber Waves, Becket, Mass., USA, pp. 56-113, 132-151, 2012.

[12] Esko, Edward and Jack, Alex, *Cool Fusion*, second edition, Amber Waves, Becket, Mass., USA, pp. 79-87, 2012.

[13] Patterson, Jeffrey, "Time to fix our nuclear waste disposal system," *Santa Monica Daily Press*, January 6, 2012.

[14] Mtingwa, Sekazi K., "Feasibility of Transmutation of Radioactive Isotopes," An International Spent Nuclear Fuel Storage Facility -- Exploring a Russian Site as a Prototype: Proceedings of an International Workshop, The National Academies Press, Washington, D.C., 2005.

[15] Michaels, Gordon E., "Partitioning and Transmutation: Making Wastes Nonradioactive," Oak Ridge National Laboratory Review, Vol. 44, No. 2, 2011.

[16] World Nuclear News, World Nuclear Association, January 11, 2012.

# INTERVIEW WITH EDWARD ESKO

*Briefly summarize the most profound transmutation results you have directly produced.*

Our results are summarized in our books, *Corking the Nuclear Genie* and *Cool Fusion*. One powerful result was the apparent transmutation of zinc and sulfur into palladium under vacuum and at relatively low temperature. About 50 to 60 amps of electric power were delivered to the vacuum tube. A significant amount of chromium and strontium also appeared. Another dramatic result was the unexpected appearance of 1,500-ppm copper from the low energy transmutation of iron and lithium. In another test, 3,500-ppm germanium appeared from the transmutation of lithium and copper.

*When you first found transmutation in an experiment, were the experiments aimed at transmutation or was it a byproduct of another process?*

In our experiments we set out to prove the transmutation hypothesis. Transmutation wasn't a byproduct; it was what we were looking for.

*What transmutation results (other than your own) seem the strongest?*

Our work was inspired by the work of George Ohsawa, which was in turn inspired by the discoveries of Louis Kervran. Michio Kushi, a student of Ohsawa's, achieved dramatic results in experiments conducted in Cambridge in the mid-1960s. It was Kushi's lectures on this topic that directly inspired our work. The 19th century experiments conducted by Sir Norman Lockyer, the discoverer of the element helium and founder of the journal *Nature,* have also provided inspiration.

*Do you think it is possible to transmute lead, uranium, thorium, etc.?*

Yes. In our book, *Corking the Nuclear Genie*, I outline experiments for the low energy transmutation of uranium-235 and plutonium-239. If my hypothesis can be proven, it could solve the problem of nuclear waste.

*Are there geological transmutations (apart from radioactivity)?*

One of our colleagues, who is from MIT, hinted to me that transmutations may be occurring deep inside the earth. Transmutation may explain the continual appearance of certain metals with other metals in common ores. It may also explain the relative abundance of iron in the earth's crust. Our experiments with converting carbon to iron may hold the key. When lightning strikes a tree, for example, a portion of the tree's carbon may be converted into iron. Lightning may also cause two atoms of oxygen in the atmosphere to fuse, forming an atom of sulfur. My own feeling is that transmutation occurs constantly throughout the universe within giant electrically charged plasma clouds. If proven, such transmutations occurring at the galactic and intergalactic level would cast serious doubt on the Big Bang.

*In the near future will it be possible to economically make noble metals by transmutation?*

It has already been done. However whether small tabletop experiments can be scaled into large industrial processes remains to be seen. In the palladium experiment, for example, two inexpensive elements—zinc and sulfur—are used to produce palladium, a much more expensive element. If the experiment can be scaled, it could be quite profitable.

*What benefits can society reap from the transmutation of elements in particular?*

Transmutation of the elements is one side of the coin that will free humanity from the twin scourges of poverty and ignorance that have plagued it from the beginning. The other side is energy. Solving the mystery of the Great Pyramid, for example, may help greatly in that regard. For that we need to study the phenomenon of buoyancy, or as Edgar Cayce put it, "the way that iron swims" in water. An ocean of air surrounds us. It might be possible to apply the principle of buoyancy to float heavy objects, like the giant blocks of the pyramid, in the air.

The principles of transmutation and unlimited energy are the principles of the universe. Understanding them will free humanity once and for all from the scourge of ignorance and the scourge of poverty.

Source: *Infinite Energy Magazine*, November-December, 2018.

## ABOUT THE AUTHOR

Edward Esko, founder of Quantum Rabbit LLC, designed the experiments outlined in *Cool Fusion, Corking the Nuclear Genie,* and *In Search of Nanonovae.* He is the founder of the Quantum Research Institute, a global initiative to further the study and application of low energy transmutation.

QuantumResearchInstitute.com.
edwardesko@gmail.com